Food 117

山羊爱苹果

Goats Love Apples

Gunter Pauli

［比］冈特·鲍利　著

［哥伦］凯瑟琳娜·巴赫　绘

颜莹莹　译

上海远东出版社

丛书编委会

主　任：田成川

副主任：闫世东　林　玉

委　员：李原原　祝真旭　曾红鹰　靳增江　史国鹏

　　　　梁雅丽　孟小红　郑循如　陈　卫　任泽林

　　　　薛　梅　朱智翔　柳志清　冯　缨　齐晓江

　　　　朱习文　毕春萍　彭　勇

特别感谢以下热心人士对童书工作的支持：

匡志强　宋小华　解　东　厉　云　李　婧　庞英元

李　阳　梁婧婧　刘　丹　冯家宝　熊彩虹　罗淑怡

旷　婉　王靖雯　廖清州　王怡然　王　征　邵　杰

陈强林　陈　果　罗　佳　闫　艳　谢　露　张修博

陈梦竹　刘　灿　李　丹　郭　雯　戴　虹

目录

Contents

Z RI Learning Initiative

一群山羊正在巴斯克地区的高地上漫步，那里风景壮丽，美景从山顶一路延绵至大海。一只燕雀看着这些饥饿的山羊走向一棵苹果树，开始用力咀嚼鲜脆的果子。

"你们喜欢吃苹果？"燕雀问道。

A herd of goats is roaming the highlands of the Basque Country, with its breath-taking views from the top of the mountains all the way to the sea. A finch is watching as the hungry goats reach an apple tree and start munching on the crisp fruit.

"Do you like to eat apples?" the finch asks.

一群山羊正在高地上漫步……

A herd of goats is roaming the highlands ...

难道你不知道其实我们会爬树吗？

Don't you know that we do climb trees?

"当然，我们很爱吃！"一只山羊回答，"在秋天我们尽情享用苹果，苹果吃完了，我们就吃梨，再之后还有栗子。"

"幸亏你们不会爬树，要不然什么都不会给我剩下了，我可真幸运！"燕雀开玩笑说。

"难道你不知道其实我们会爬树吗？我们不只是吃那些掉落在地上的果子。你停在树枝上的时候可要当心了。"山羊警告说。

"You bet, we do!" responds one of the goats. "We feast on these in the fall, and eat all we can. When the apples are gone, we will enjoy pears, and after that feast on chestnuts."
"How lucky am I that you cannot climb trees, or there would have been nothing left for me," Finch jokes.
"Don't you know that we do climb trees? We don't only eat what drops to the ground. Watch out for us along those branches," Goat warns him.

"我得承认，我从来都不知道山羊吃苹果、梨，还有栗子！"燕雀坦白说。

"你真的认为我们只吃草吗？那得多无聊哇。我们有时还是想尝试点不一样的，一些真正美味又健康的食物。"山羊说。

"是啊，只吃一种食物我也会不开心。但我不能只靠苹果为食，它们的生长季太短了。"燕雀指出。

"I must confess, I never knew that goats eat apples, pears or chestnuts!" Finch admits.

"Did you really expect us to eat only grass? How boring that would be. No, now and then, we want to eat something different, something really tasty and healthy," Goat says.

"Well, I would not be happy eating only one type of food either. But I cannot live off apples only. The season is too short," Finch points out.

......一些真正美味又健康的食物......

... something really tasty and healthy ...

蠕虫和蛆虫……

Worms and maggots ...

"那你还吃些什么呢？"山羊问道。

"春天吃一些蠕虫和蛆虫，一年大部分时候吃各类种子。你要知道，冬天对我们来说很难熬，我需要飞到港口那边看看有什么东西剩的。"燕雀解释道。

"So what else do you eat?" Goat wants to know.
"Worms and maggots in the spring. Seeds of all types most of the year. And you know, for us the going gets tough in winter. Then I need to fly down to the harbour and look for left-overs," Finch explains.

山羊笑着回答：“我们哪儿都不用去，农民会把食物给我们送过来。”

　　“你的意思是，你们能够让农民把食物带给你们，而不用自己去寻找？”燕雀问道。

　　“的确是这样！”山羊肯定地说。

Goat smiles and replies: "We do not have to go anywhere, the farmer brings us our food."
"You mean to say you have trained the farmer to bring food to you, instead of you having to go look for it?" Finch asks.
"Indeed!" Goat affirms.

......农民会把食物给我们送过来。

... the farmer brings us our food.

······暖气流。

...warm air currents.

"你可得教教我怎么做！实不相瞒，虽然飞到山下很容易，但再飞回来对我来说却很费劲。我常常必须等到有暖气流的时候才行。"

"想让人类喂养你，你必须对他们有所回报。"

"You have to teach me how to do that! Flying down these hills is easy enough, but I tell you, getting back up here takes it out of me. I usually have to wait for one of these thermals, these warm air currents."

"To get people to feed you, you have to give them something in return."

"那你给他们的回报是什么呢？"燕雀问。

"羊奶。我们产羊奶……"

"那不是意味着你的孩子就没奶喝了？"

"噢，当我有了小孩子，我的孩子喝完奶后，农民会持续给我挤奶，这样就算我的孩子不再需要喝奶了，我也会连续数月甚至数年产奶。"山羊解释道。

"And what is it that you are offering them?" Finch wants to know.

"Milk. We produce milk ... "

"Does that not mean your kids don't get any?"

"Well, when I have a little one, and my kid has suckled, the farmer keeps on milking me. In this way I produce milk for months and sometimes even years after my kids no longer need it," Goat explains.

羊奶。我们产羊奶……

Milk. We produce milk …

奶酪和冰激凌......

Cheese and ice cream ...

"太有趣了！农民拿你们的奶做什么？自己喝吗？"

"哦，不，喝那么多奶他会消化不良。最初他用多余的奶做奶酪，现在他会用来做冰激凌了，他为此感到很高兴。孩子们很喜欢冰激凌。"

"How interesting! And what does the farmer do with all your milk? Drink it himself?"

"Oh no, he would get indigestion from drinking so much milk. At first he made cheese from the extra milk. And now he is excited about making his own ice cream too. Kids love it."

"所以你的孩子们很高兴，他的孩子们也很高兴。这一定让他也很高兴！"

"他喜出望外，因为把奶酪和冰激凌卖出去他还可以挣到钱。"山羊骄傲地说。

"我在想，这种冰激凌尝起来会不会有苹果味呢？"

……这仅仅是开始！……

"So your kids are happy – and his kids too. That must make him very happy as well!"

"He is overjoyed, as this way he is able to earn more – by selling cheese as well as ice cream," Goat says proudly.

"Now I wonder, will this ice cream taste naturally of apples?"

... AND IT HAS ONLY JUST BEGUN! ...

......这仅仅是开始！......

...AND IT HAS ONLY JUST BEGUN! ...

Did You Know?

你知道吗?

山羊是大约 9 000 年前最先被驯化的动物之一。在英语里,母山羊也被称为 nanny,公山羊被称作 buck,小山羊被称为 kid。

Goats were one of the first animals to be domesticated about 9,000 years ago. Female goats are called nannies, males are bucks, and baby goats are called kids.

There are approximately one billion goats farmed, the largest meat supply on earth. Goat wool is called mohair. Special breeds of goat that survive months of ice-cold winter temperatures produce ultra-soft cashmere.

全球大约有 10 亿只山羊被畜养,它们是最大的肉类来源。山羊的毛叫作马海毛。能在极寒的冬天存活数月的山羊品种,可以产出极其柔软的羊绒。

山羊是反刍动物，所以有4个胃。但是山羊刚出生的时候，它们的瘤胃（第一个胃）还未充分发育，所以它们消化食物的时候就好像只有一个胃。山羊奶是碱性的，20 分钟就可以消化，牛奶是酸性的，需要一个小时才能消化。

Goats are ruminants, thus have four stomach compartments. But when goats are born, their rumen is underdeveloped, so they digest as if there is only one stomach. Goat's milk, which is alkaline, is digested in 20 minutes, whereas acidic cow's milk takes an hour.

山羊是高地动物，善于攀岩；它们能够爬到树顶寻找最爱吃的果实。它们的 340 度全景视野帮助它们发现各处的食物。

Goats are mountain animals and climb very well; they have been known to climb to the top of trees to find their favourite fruit. Their 340-degree panoramic vision helps them to find food everywhere.

山羊与人类关系紧密，对待牧羊人十分友善。通常，生活在 50 头左右的小型羊群里的羊的产奶量要比生活在大型羊群中的高 10%—15%。

Goats develop a bond with humans and appreciate the care given by the herder. Goats in small herds of up to 50 tend to produce 10%-15% more milk than goats that live in larger herds.

燕雀是地球上最小的鸟类之一，而雀科是鸟类中最大的家族之一。人类的出现容易使雀类感到紧张。

Finches are one of the smallest birds on Earth, and the finch family is one of the largest bird families. Finches are stressed by the presence of humans.

Finches are native to Europe and were introduced to the Americas as cage birds by pet dealers. Today, finches are the most common garden birds in cities like New York and Los Angeles.

燕雀是欧洲本土鸟类，被宠物商作为笼鸟引入美洲。今天，在纽约、洛杉矶等城市，燕雀是最常见的花园鸟类。

Finches have made a great contribution to science, inspiring Darwin to formulate his Theory of Evolution. Finches are also inspiring new science as they seem to have their own grammatical rules in songs, and have been proven to have a unique sense of smell.

燕雀对于科学的贡献巨大，达尔文的进化论灵感就来自燕雀。燕雀对新科学也有所启发，它们在歌唱时似乎有自己的语法规则，并已被证实拥有独特的嗅觉。

Think About It

想一想

Do you think it is a good idea to own a goat or a finch as a pet? If you do own a pet, should you have one or more?

你觉得养一只山羊或者燕雀当宠物是个好主意吗？如果你有宠物，你会养一只还是好几只？

Goats in small groups with a caring farmer herding them produce more milk. What would you do to have goats increase their milk production?

一小群羊再加上一个细心照料的农民，会使山羊的产奶量增加。如果想增加山羊的产奶量，你会怎么做？

How do you think the goats feel now that the farmer brings them food instead of them having to forage for food? For one thing, they do not have to climb trees anymore ...

山羊由农民饲养，不用自己出去寻找食物，你觉得山羊会有什么感受？一方面，它们再也不用爬树了……

Do you think one makes more money selling cheese than ice cream?

你认为卖奶酪会比卖冰激凌更赚钱吗？

Do It Yourself!

自己动手！

We are going to make ice cream with goat's milk! You will need: 4 cups of goat's milk, 1 cup panela (unrefined whole cane sugar) instead of sugar; ¾ teaspoon vanilla essence, the zest of a lemon, a pinch of salt, and 4 egg yolks. First warm the milk, panela, vanilla and salt in a saucepan until the panela dissolves. Beat and temper the egg yolks (raise the temperature gradually), add it to the milk mixture. Now add the lemon zest. Boil for 2 minutes, stirring constantly. Remove from heat and strain. Refrigerate, but continue to stir until completely chilled, avoiding the formation of crystals.

我们将用羊奶做冰激凌！你需要准备：4 杯羊奶，1 杯红砂糖（甘蔗的原榨汁）以代替白糖，3/4 茶匙香草精，柠檬皮，少量盐，以及 4 个蛋黄。首先将奶、红砂糖、香草精和盐倒入锅中加热直到红砂糖融化。将蛋黄打散，倒入奶中，逐渐升温加热。现在加入柠檬皮。煮沸两分钟，持续搅拌。关火，过滤。冷却的同时继续搅拌，直到完全凝固，注意防止过程中形成结晶。

27

学科知识
Academic Knowledge

生物学	解剖山羊，会发现它的瞳孔是狭长的而不是圆的；反刍动物成年与童年时的消化过程；山羊是食草动物，燕雀是杂食动物；超出哺乳期需求的产奶能力；每只雄燕雀的歌声都是不同的，但是家族成员间会有相似性；雄性斑胸草雀的歌声习得来源于周围环境。
化 学	羊奶是碱性的，牛奶是酸性的；羊奶的胆固醇含量低，钙、磷、维生素A含量高。
物 理	利用阻力攀爬；利用气流移动。
工程学	奶业的工业化和大规模制造奶酪；畜牧业规模的扩大。
经济学	山羊产肉、奶、粪便及纤维（马海毛及羊绒，现在被人工合成纤维取代）；在奶业里冰激凌行业利润率高，比奶酪行业现金流动快；小规模放牧比大规模放牧更高产——小而美，盈利也更可观。
伦理学	有所回报与不求回报。
历 史	山羊是最早被驯化的动物之一；从采集狩猎到农业社会的转变是历史转折点；用山羊做祭品的传统；达尔文进化论的形成是建立在对燕雀的研究基础上的。
地 理	巴斯克地区高地；加拉帕戈斯群岛燕雀的多样性。
数 学	如何计算阻力、功和功率；计算奶酪、奶、冰激凌的产量。
生活方式	吃应季食物是保持健康的方法之一。
社会学	群体归属感；在集体中承认自己知识不足的信心。
心理学	昏厥作为山羊的一种防御机制；去探索与发现的强烈好奇心；与孤独相关的抑郁产生的影响；通过关爱提高产量。
系统论	山羊全身都是宝，包括粪便（可作为燃料）和骨头。

教师与家长指南

情感智慧
Emotional Intelligence

燕 雀

燕雀具有同理心，提出许多相关问题。与山羊相处使他感到舒适，所以会开玩笑。他承认自己并不知道山羊会爬树，显示出他有自信。他指出冬天他面临压力，不得不飞得很远去找食物，并对山羊与牧羊人之间的亲密关系感到很吃惊，想知道动物是如何使人类为其提供食物的。他认为山羊的奶应该给她自己的孩子喝，而不是给人类。他迫切地想要了解事物背后的逻辑，并认为自己有权多问问题。他能够全面看待问题，指出牧羊人很开心，并总结出这是一个对双方都有利的安排。

山 羊

山羊热情积极地回答燕雀的问题，展示出她的确定感和适应能力。她对待燕雀十分友好，并不介意燕雀不知道她会爬树。山羊不断分享新鲜知识，表现出对燕雀的兴趣与友善，想要了解燕雀的饮食习惯。山羊自知生活舒适，农民在食物短缺的冬季也会满足山羊的所有需求，这令山羊对她的生活感到满意。山羊认为这是一个公平交易，农民提供食物，得到羊奶。山羊很开放，乐于讨论，并分享了如何在哺乳期结束后继续产奶的细节。山羊知道将奶加工成孩子们爱吃的冰激凌会给农民带来收益。

艺术
The Arts

斑胸草雀是一种常见的燕雀，研究它的羽毛，观察斑马效应。现在让我们用颜料和铅笔重现斑马效应。使用颜料画出深浅交替的图案，第一笔涂得黑一些、重一些，第二笔浅一些，深色笔画中间留出白色空间。这样就创作出一个黑白主题的图案。尝试不同风格，看看哪一幅作品可以作为墙壁装饰，哪一幅与斑胸草雀的外观最接近。

29

思维拓展
Systems: Making the Connections

地球上的生命都遵循季节。每一个季节都有其特有的产物，提供当季所需的食物与营养。但是，食物的全球化使同一种食物全年可供应，让我们对食物的自然周期的感知变得麻木。受到标准化与追求低成本的规模经济驱使，全球的餐饮业趋于提供一份稳定和相似的菜单，这同时也意味着文化、传统、气候与地理适应性的逐渐丧失。这种耕作和生产方式缓解了冬季的压力，填补了两个收获季之间的供给短缺，却导致供应链的简单化。除了受减少作物的数量和种类、只种植有限数量的作物影响，作物产量还受到不断增产的压力与提高产量的需求的影响。拥有一家有着1万头牛和5万头猪的饲养场已不再是什么新鲜事。事实上，由系统来模拟这些规模化操作的需求与日俱增；通过系统，食品的设计和产量变得可以预测。牧羊成为一类有趣的实验。人们普遍认为牛是肉类的主要来源，但世界肉类市场实际上由羊主导。山羊是有感情的动物，能够很好地回应农民的关爱。值得注意的是，在小规模羊群（50只以内）中喂养的山羊产奶量更多。当山羊超过一定数量，农民就很难关注到每一只羊，给每只羊都起名字并记住它们。小规模放牧，牧羊人就有办法记住每一只羊。这种方法遵循了经济学家舒马赫的逻辑，他表示"小即是美"，而现在，小的也是更高效的。这对我们所有人来说都是一个新的课题。

动手能力
Capacity to Implement

你生活在一个四季分明的地区吗？不管生活在哪里，总有水果和蔬菜是无法全年供应的；每日的新鲜供应总会有中断。所以与其从全球各地采购，不如让我们列出一年中每个月能收获的当地农作物清单。现在，说明为什么我们应该更多地去买卖和食用当地的几种应季作物而不是去世界各地采购各种各样的全年可供应农作物。与你的朋友和家人分享你的观点和想法，告诉他们你认为这么做有道理的原因，以及如何从健康的角度出发去行动。

故事灵感来自
This Fable Is Inspired by

哈维尔·莫拉莱斯
Javier Morales

哈维尔·莫拉莱斯是一名农学家，是西班牙加那利群岛耶罗岛人。他常常跟山羊打交道，并观察到给予动物关爱会对奶产量造成影响。哈维尔当选为加那利群岛地区议员，并一直致力于孤岛的经济转型工作。虽然许多传统经济学家将哈维尔和他的同事们在岛上的工作视作向自给自足的封闭经济的回归，哈维尔却认为这种尝试使下一代明确了面临的挑战，培养了一种能在现代社会边缘生存下去的新竞争力。哈维尔在岛上成功推广了一系列合作模式和生产型企业，并创建了一个维持水和能源自给自足的战略平台。现在他正在从事一项新的电动车计划，使用的电力将完全依靠岛上的可再生能源。

图书在版编目（CIP）数据

冈特生态童书.第四辑:修订版:全36册:汉英对照 /
（比）冈特·鲍利著;（哥伦）凯瑟琳娜·巴赫绘;
何家振等译.—上海:上海远东出版社,2023
书名原文:Gunter's Fables
ISBN 978-7-5476-1931-5

Ⅰ.①冈… Ⅱ.①冈…②凯…③何… Ⅲ.①生态环
境−环境保护−儿童读物—汉、英 Ⅳ.①X171.1-49

中国国家版本馆CIP数据核字（2023）第120983号
著作权合同登记号图字09-2023-0612号

策　　划　张　蓉
责任编辑　张君钦
封面设计　魏　来　李　廉

冈特生态童书
山羊爱苹果
[比]冈特·鲍利　著
[哥伦]凯瑟琳娜·巴赫　绘
颜莹莹　译

记得要和身边的小朋友分享环保知识哦！
八喜冰淇淋祝你成为环保小使者！